STUDIES IN CONTRACTUAL CLAIMS 4

Contractual claims under the ICE Conditions of Contract

BY W.E.I. ARMSTRONG, OBE, TD, DL, MEng, CEng, FICE, FIMechE, FCIArb, FCIOB

The Chartered Institute of Building
Englemere, Kings Ride, Ascot, Berkshire

CONTENTS

1. Introduction

This paper forms part of the series on contractual claims issued by The Chartered Institute of Building[1,2,3], to help provide those two successful ingredients of any successful claim; experience and knowledge.

The series draws on the acknowledged experience of experts such as Burke and Wood. This paper is concerned with the ICE Conditions of Contract.

Much of the guidance given here — particularly in the opening section — is also applicable to other forms of contract, though great care must always be exercised in applying principles from one contract to another.

Detailed information on tendering and construction items is given in the paper by Wood[3]. It is not repeated here because methods of tendering, including rate build-up, distribution of overheads and preliminaries vary more with the type of work and size of firm than with the conditions of contract.

CLAIMS FOR ENTITLEMENT

Claims for entitlement under the ICE Forms of Contract are made to recoup reasonable costs of the contractor caused through events which are the responsibility of the employer. The ICE Conditions of Contract, 5th Edition, give considerable protection to both the employer and the contractor. The vast number of claims for entitlement, when properly presented along the lines discussed, are settled at a low key, satisfactorily and reasonably speedily.

Those which are not so settled may run for months or even years and involve extended discussion, conciliation, often by a third party representing the contractor's case to the engineer, and eventually, arbitration.

It is not unusual for an engineer's subordinate to turn down a claim when presented and re-presented two or three times. A contractor with a sound claim for entitlement must not be put off course by such action. The subordinate's advice may be wrong. The contractor may benefit by asking to see the engineer. Whatever happens, the contractor must pursue the claim resolutely, appreciating the situation frequently as the claim process progresses. Many contractors go out of business through not pursuing their claims resolutely and with reasonable speed.

Of concern is the time taken for contractors to receive certain payments, particularly where the employer is acting for another authority, and the leaching away of some of the engineer's power by others.

The engineer's authority to use his personal judgement to apply the contract fairly is absolutely essential for the smooth working of the contract. Clause 66, with arbitration in the background, is the present safeguard.

The number of cases where an arbitrator is appointed is relatively few and the number actually reaching the arbitration hearing even less.

This is a great credit to the work done by all parties and the co-operation, generally given, particularly by the engineers to the contracts, whose lot is not always an easy one.

Engineers often look upon claims from contractors as an attempt to recoup an avoidable loss at the employer's expense. This is due largely to so many contractors' claims being presented badly.

The claim has to be presented in contractual terms and must show why the employer is responsible for the cost involved. The fact that the engineer may be responsible directly for the cost should not influence his judgement.

The entitlement within the contract must first be proved, the extension of time involved decided, and the entitlement quantified.

It must be remembered that a claim is the demand for reimbursement of the cost incurred by the contractor on behalf of the employer. Costs in this context include overhead costs, whether on or off the site, except where the contrary is expressly stated.

ICE CONDITIONS OF CONTRACT 5th EDITION

The Conditions of Contract and Forms of Tender, Agreement and Bond for use in connection with Works of Civil Engineering Construction, Fifth Edition (issued June 1973)(revised January 1979)(reprinted January 1986) approved by the Institution of Civil Engineers, the Federation of Civil Engineering Contractors and the Association of Consulting Engineers is in this paper referred to as the '5th Edition'. The type of contract irrespective of edition, is referred to as the 'ICE contract'.

The January 1986 Reprint includes amendments covered by guidance notes four, five and six issued May 1982, September 1983 and June 1985 covering clauses 60(6), 34, 66 and 67 and has minor amendments to clauses 1(1)(c), 30(1) the form of tender, and sub clause 4(c) of the contract price fluctuation clause to clarify and bring them up to date.

Works of a similar nature may have different Conditions of Contract, depending on the employer and his advisers' wishes. The contractor's price will normally differ for the same works under different conditions or, for that matter, different editions or variations of an established contract form. The exact wording of the form of contract used must, therefore, be studied in a claim for entitlement, including any variations.

Contracts at present being signed for works of civil engineering construction are normally based on the 5th edition and this paper is based on that edition except where indicated.

The 5th edition is often considered by engineers to be favourable to contractors. This is not true, because a tender price reflects the conditions and the protection given by the Conditions of Contract to both parties to the contract.

4th EDITION OF THE CONDITIONS OF CONTRACT

The 4th edition of the ICE contract (issued January 1955) is rarely used now but claims for entitlement based on this edition will continue to be made for a few years. As time progressed additions were made to the 4th edition. These are listed and described on the front cover of the 5th edition.

Many of the claims situations differ between the 4th and 5th editions as does the contractor's entitlement. So the two editions must not be confused.

FORM OF SUB-CONTRACT

The form of sub-contract designed for use in conjunction with the ICE General Conditions of Contract is referred to in this paper as the 'sub-contract form'. It is used frequently for sub-contracts — particularly nominated sub-contracts — but it is by no means universal. The exact wording of the sub-contract used must be studied in all cases.

Particular care must be taken to determine whether the sub-contractor is nominated or domestic. Frequently, a sub-contractor will consider he is nominated when, in contractual terms, he is a domestic sub-contractor.

Care must also be taken to establish whether the sub-contractor made an offer which was accepted by the contractor, or whether the contractor gave an order which was accepted by the sub-contractor.

Many contractors find themselves caught in the middle when the sub-contract conditions are not co-ordinated with the terms of the main contract. In this situation the main contractor may find himself responsible for something which he has neither priced nor anticipated.

Some of the more general problems arising with sub-contractors are dealt with later.

DISCOVERY OF A CLAIM SITUATION

Making a loss does not imply that the contractor has grounds for making a claim. When profits do not meet expectation the contractor should examine his progress against the programme envisaged in the tender to see where it differs. Some problems may be of his own making or responsibility, whereas others will be those of the employer and his representatives. Responsibility should be indicated by the contract, although some 'grey areas' will exist. Aspects normally associated with the contractor may, in fact, be the employer's responsibility since his actions or those of his representatives may have forced the contractor into extra cost.

It is often difficult to discover where the extra cost has arisen but little progress can be made until this is solved. However, once identified, examination will determine whether the contractor has an entitlement for reimbursement and whether a claim, or claims, can be made.

The progress plotted against the programme will often indicate delays, out of sequence working, disruption and disorganisation. Examination of the cause may result in a justifiable claim being made. Similarly, an examination of trade labour expended against return may show excess cost, as might plant and preliminaries. The cause may be one of low outputs of labour and plant; the reason for this must also be determined.

If the contract is running late an extension of time for completion will normally be required. This should be examined in a similar manner. Some extensions of time carry money entitlement whereas others do not.

2. Foundations for a successful claim

Having discovered the existence of a claims situation, the following are the important foundations upon which a successful claim is based.

Entitlement
Too many claims are based on the hope that the engineer's sympathy will suffice. There may be grounds for an ex-gratia claim (see section on types of claim) but they are not granted frequently; this is particularly true of government and local government contracts. An entitlement within the contract must be shown, stated and proved. It is usually necessary to state the clause or clauses under which the claim is made.

Employer's responsibility
Most clauses require the employer's responsibility to be shown. This includes action or inaction of the engineer and others, but it must be shown clearly.

Notification of the claim
Where the terms of the contract require notification of a claim, it is essential that this is given and in accordance with any time requirements. Failure to do so may not invalidate the claim, but the cost and expense is more likely to be disputed and reduced to the contractor's disadvantage.

Extension of time
An extension of time must be requested where appropriate. It may be needed to reduce liquidated damages as, for example, in the case of adverse weather conditions where additional cost and expense are normally the responsibility of the contractor. Equally, an extension may be needed as a basis for a claim for entitlement to increased preliminaries and overheads, or extra labour and plant. The contractor must be careful that he is not given excessive extensions of time on items which carry no money entitlement and which may prejudice his genuine claims for entitlement later.

An extension of time is not always applicable to a claim for entitlement. The claim may be for an item not on the critical path, the time may be regained at extra cost and expense, the contract may have been staffed and programmed to finish early, or other reasons may apply. Where the contract states that an extension of time must be requested, then notice must be given in accordance with the appropriate clauses. The actual time involved will be determined subsequently.

Cost records
Frequently when the engineer is notified of a claim he is required to state what records he requires to be maintained. In other situations he will take the initiative, but in any case records must be kept. Records must be capable of establishing the extra cost and expense incurred.

Accuracy
Many claims fail because the records are shown to be inadequate or incorrect. In some cases the engineer may award an amount less than that entitled to the contractor.

'Manufactured' figures rarely survive close examination. At arbitration and in the Courts they can prove expensive, both from the costs awarded and the loss of awards on claims in general.

Records must be accurate and consistent. Substantiation by the engineer's representative is important. Photographs are particularly useful as supporting evidence.

TYPES OF CLAIM

The main types of claim are:

i) *Star rates and star items*
Frequently, an engineer may prefer to reimburse the contractor's costs and expense by altering rates generally known as 'star rates' in the bills of quantities or 'star items' ie, new items with a different description from that existing: such items are normally less contentious. It may be expedient, before the claim is submitted, to discuss with the engineer the way he wishes the claim to be submitted. He may even be in a position, contractually, to award star rates or star items without a formal claim for entitlement as, for example, under Clause 52(1) or Clause 56(2).

ii) *Contractual claims*
Contractual claims, excluding amounts paid on star rates and star items, will require to be built up and presented as described in the following sections. Compilation takes time. It is, therefore, essential that the engineer knows the claim is being produced and that

reasonable attention and urgency is given to limit the delay in production.

iii) *Ex-gratia claims*
Ex-gratia claims occur more commonly with industrial and private clients where the contractor has suffered loss and expense not covered by the contract but for which the client considers he has a moral responsibility or responsibility in equity and frequently in tort. In many such cases the contract does not cover all the situations which arise and this method is used as an acceptable way of settling the matter. This is particularly so where the goodwill on both sides is important for future co-operation.

With government and local government contracts such arrangements normally require high level authority. They should not be relied upon as a means of recouping losses.

DISAGREEMENT ON RATES NOTIFIED
If the contractor intends to claim a higher rate or price than the one notified to him by the engineer under Clause 52(1) and (2) or Clause 56(2) he must, within 28 days after such notification, give notice in writing of his intention to the engineer in accordance with Clause 52(4)(a). This is most important.

INCREASED PRELIMINARIES, OVERHEADS, FINANCIAL CHARGES AND SIMILAR
Preliminaries, overheads etc, are distributed over the priced bill of quantities in a number of ways, depending on the size of the job, the size of the contractor and the facilities available at the time of tender for rapid computation.

The preliminary items in the preliminaries bill do not cover all preliminaries and site costs, let alone the other overheads. Depending on how the bill is priced, the amount for preliminaries, overheads, financial charges and similar costs will be recouped by additions to the preliminaries bill, the rates and PC sums and the balancing item. The amount involved when extracted (£T) will have been based partially on the intended length of the contract, partially on the quoted value at the tender stage and partially on a mixture of time and value. An example is given in Appendix I on determining the amount built into a tender for preliminaries, overheads and profit.

TIME INCREASE
The contract may start late and may be extended in time. When the employer is responsible for the delay, the contractor is entitled to reimbursement for both these delays on the time element and part or whole of the mixed element. It may be based on all of the £T.

The contractor will have submitted a tender based on a certain proportional reimbursement of overheads each week. If the job is late in starting or delayed he cannot obtain other remunerative work because he has allocated his resources and finance to that job. Consequently, the amount to be claimed for this will be:

$$\frac{a + b}{c} \times d$$

where a = time element amount
b = part or whole of the time/value element
c = number of weeks intended for the contract
d = number of weeks of the respective delays

It may be that the full amount £T calculated should appear above the line; this will depend on circumstances. The amount per week may differ for the delay which occurs before the contract starts and for the delay which occurs during the contract. Some items may increase more than the pro rata rate; this is particularly true of the financing element and adjustments may be required. Similarly, an allowance will need to be made for the effect of inflation on the total £T, whether the job is firm price or not. This is a cost that the employer and his representatives have imposed on the contractor.

Frequently, the amount £T is not distributed evenly. This can be seen from the programme, or other documents, or the way the bill is priced. It occurs particularly with sectional completion or plant installation and must be allowed for in the claim.

If the number of weeks intended for the contract by the contractor at the tender stage is less than the time stated for completion in the Appendix, then it is an advantage to make this clear at the time of tender.

VALUE INCREASE
Where the value goes up, a similar calculation is necessary. There may be some contribution from the additional items in the final account.

OUT OF SEQUENCE WORKING, DELAY, DISORGANISATION AND DISRUPTION
When there is out of sequence working, delay, disorganisation and disruption on a contract, extra staff may be needed and even more staff may be required to bring the job back on time if so required. Staff labour for short periods may cost

more than the cost of normal quota of staff. If the job goes on longer, the cost of the staff in the extended period will be greater than the average due to inflation. Extra huts, cars, small tools, etc, may be required.

All these must be costed and allowed. Some may be included in the time and value increases referred to above but the amounts concerned have to be separated and sorted out to reimburse the contractor fairly.

EXTRA PLANT AND LABOUR COSTS

The contract will normally be scaled on plant and labour according to the programme on a set number of weeks for the contract. Delay, out of sequence working, disorganisation and disruption will normally increase both the plant and labour elements. If the job takes twice as long for the same amount of work, twice the plant cost (or more) may occur. Labour will normally work less productively, strikes and overtime may be involved and a non-productive bonus may have to be paid, all of which must be costed. The Engineer's agreement at the time helps.

The contractor will be able to establish his additional financial commitment from his labour and plant costs, compared with those intended. A percentage of labour expended may be adjudged, eg, 30% depending on the circumstances.

OTHER ADDITIONAL, REASONABLE COSTS

Increased costs over those allowed

In tendering on a firm price contract an addition is made for increased costs based on the tender programme and assessments of cost increase per annum (or during the contract period). If the time for completion is extended, due to the employer or his representatives' action or inaction, or even if parts of the job are disrupted or delayed, extra increased costs over those allowed occur which must be claimed. This may apply even when the contract is completed on time or before time. These extra costs are the 'reasonable costs' which would not otherwise have occurred.

Fluctuations and formula methods

A contractor, when tendering, allows for the fact that with the fluctuations clause, he will receive a proportion only of his increased costs. He will also allow for the extra costs of staff and the like. With the formula method there is a small amount unreclaimable and this still has to be allowed for in the tender build-up.

The same remarks in the previous section apply to the increase in the unreclaimable parts which results from any action or inaction which is the responsibility of the employer. This has to be calculated and included in the 'reasonable costs' in the same way.

Winter working

Where the employer's action or inaction increases the amount of winter working, the reasonable extra cost of this needs to be calculated and added in.

Weather and labour problems

Where the employer's action or inaction results in the contractor having to work in bad weather and/or causes labour problems which would not otherwise have occurred, then the extra costs become an additional reasonable cost. Disruption or disorganisation may be just as important as delay, particularly with regard to labour problems.

Additional interest on overdraft

Excluding the higher rate of interest which may be required for finance when delay takes place, the financing charge may increase due to the different deployment of the cash flow. Any extra cost should be added in as a reasonable cost, including the extra cost of financing the retention.

SELECTING THE RIGHT CLAUSE ON WHICH TO BASE A CLAIM

Selecting the right clause is of great importance. If in doubt it is better to select more than one clause and leave the final decision to the engineer — quoting sub-clauses unnecessarily may restrict the engineer. He, too, has to consider the contractual implication of his choice and may wish to have room to manoeuvre. Auditors and others will probably vet his choice.

MAIN CLAIMS CLAUSES

The major clauses under which claims are made for reasonable extra costs are listed in Table 1. There are other, less commonly used, clauses which apply to particular cases such as frustration and war.

A complete study of all the documents is always essential. This is particularly so where amendments, alterations and additions to the documents have been made, especially to Clause 12. These may force the contractor to use other clauses in pursuing his claims.

DUTIES OF THE ENGINEER

The duties of the engineer are given in general

terms throughout the contract and specifically in some clauses (see later). He will be more sympathetic to some claims clauses than others, but whichever clause is selected it is essential to keep him informed.

With large authorities, the engineer named in the contract is often remote, or insufficiently concerned with the contract, and may not know of its existence. Too much is often left to subordinates in the engineer's office whose ability varies. Such persons are not the engineer named in the contract and this must be remembered.

GIVING OF NOTICE
The period for giving notice may affect the choice of clause. Where the giving of notice is specified this must be complied with, otherwise at a later date the entitlement may be lost or reduced. Alternatively, the contractor may have to use another clause to claim his entitlement.

The engineer may have power to make an award from the information he may subsequently glean, when the notice has been given late or not at all. The award in this case may be on the low side.

A brief memorandum giving the notice is normally sufficient, fuller details being provided later. The briefer the notice the better, since the contractor may only have part of the information at this stage and later he will probably be held to what he has said. Notifications must be in writing in accordance with the contract clauses selected, both time and cost being claimed as appropriate.

LEGAL WORKS OF REFERENCE AND CASES STATED
In selecting the clauses to use, regard must be had to past judgements, and works of reference, such as Hudson[4] and Abrahamson[5].

FAILURE TO NOTIFY AND INSUFFICIENT RECORDS
Failure to notify a claim will often force a contractor to claim his entitlement under a less favourable clause. Insufficient records may limit his choice or force him to change later. Good records, verified by the engineer's representatives, are of utmost importance and they must be preserved after contract completion until all settlements have been made. Where the engineer's representatives will not verify, this should be recorded in writing to the engineer.

MINDS OF PARTIES AND INFERRED TERMS
When selecting a clause, consideration must be given to what was in the minds of the parties at the time of tender and what terms might be inferred, all within the context of the contract.

ALTERATIONS TO THE CONTRACT CLAUSES
It is essential to study the alterations, additions and deletions to the contract clauses. Changes may have been made because of inside knowledge of the employer or his representatives or because the 'Employer' or his representatives have had to 'pay up' before. The tender then becomes based on these alterations to the Conditions of Contract. Such modifications are to be deprecated but they must be examined. Their effect on a claim for entitlement can be quite devastating on the claimed reasonable costs.

At the tender stage all the employer's alterations must be clearly marked by the contractor. Frequently, amended paragraphs of the Conditions of Contract are reproduced with only one or two words deleted or added and it is often difficult to spot the changes. The reason for any change must be sought and allowed for in the price. It may, on occasions, even be necessary to return the documents. Quite small changes can result in an unreasonable and unacceptable risk to the contractor.

Particular attention is drawn to Clause 57 laying down the method of measurement. A different method of measurement, or a variation from the SMM, not only affects the tender price inserted in the bill but affects a contractor's entitlement at a later date and the method of presenting his claim.

Contract clause 1(1)(f) states that 'specification' means 'the specification referred to in the tender and any modification thereof or addition thereto as may from time to time be furnished or approved in writing by the Engineer'.

All changes in the specification must be checked with care, since knowledge of the operative specification is vital both at the tendering stage and on presenting a claim.

EFFECT OF OTHER DOCUMENTS
Any claim for entitlement must take into account the provisions of the following documents;

i) *Bill of quantities*
The bill of quantities may attempt, but not necessarily succeed, in incorporating inferred terms into the bill descriptions. The wording of the bill must be examined closely and, if

Table 1 Major contract clauses under which claims may be made

Clause	Description
7(3)	Delay in issue of drawings and instructions.
12	Adverse physical conditions and artificial obstructions.
13 (3)	Clarification of documents and engineer's instructions.
14 (6)	Delay in approval of methods of construction proposed, and alterations.
18	Boreholes and exploratory excavations.
22 (2)	Employer's indemnity on damage to persons and property.
26 (1) and (2)	Notices and fees and conforming with statutes.
27 (6)	Delay due to variations involving emergency and other works.
30 (3)	Employer's indemnity on bridges and roads.
31 (2)	Delay due to other contractors employed by the employer.
36 (3)	Cost of tests.
38 (2)	Cost of uncovering and making openings.
40 (1)	Suspension of work.
42 (1)	Late possession.
49 (3)	Work during maintenance period.
50	Searches required by the engineer.
51 & 52	Alterations, additions, omissions and NOTICE OF CLAIMS.
55 (2)	Correction of certain errors in the bills of quantities.
56 (2)	Increase in rate with change in quantity.
58	Provisional and PC sums.
59A (3) (b)	Direction by the engineer on nominated sub-contractors.
59B (4) & (6)	Delay and extra cost on forfeiture of a nominated sub-contract.
60 (3)	Final account claims.
60 (6)	Interest on overdue payments.

need be, challenged. Incorrect descriptions, ambiguities, different quantities and varied circumstances may result in expense to the contractor which will need to be reimbursed.

ii) *Drawings*
It is essential to see all the drawings which are available at tender stage. A record should be made of them and attention given to studying all notes on drawings, asking for further drawings in good time and keeping a site register of the date of arrival of all drawings and modification drawings, including sketches from the resident engineer.

All too often, drawings containing information which is not given elsewhere, arrive late or are varied late. For example, the shuttering may have been produced when a requirement for different shuttering is received. This will involve extra cost.

It is essential to ensure that the engineer knows explicitly, in writing, when drawings are required, whether based on the programme, letters, site minutes or a combination of these. Many altered drawings bear no date and the contractor can lose money because he had no record of when the late information, or alteration, arrived.

iii) *Soil mechanics surveys and reports*
Frequently, the tender documents draw attention to the availability of soil mechanics surveys and reports. These should always be examined at the tender stage and a note kept of the name of the employer's representative concerned. Contractors frequently jeopardise their interest by their failure to examine these documents at the tender stage.

iv) *Other surveys and reports and other documents*
Other surveys and reports should be examined in a similar manner. To ask for information from the employer's representative over the telephone at the time of tender can create a problem subsequently when the engineer states that if the contractor had visited his office he would have shown him the information and amplified this if necessary.

v) *Contractor's tender letter*
It is essential to incorporate any contractor's tender letter into the tender by adding, on the Form of Tender, such words as '......... and our letter dated ....'. The letter may otherwise be torn off the documents and lost.

vi) *Correspondence*
Correspondence should always be examined when preparing a claim. Oral instructions must be confirmed in writing in accordance with Clause 51 (2). Any disputations should be acknowledged. The site correspondence file should not be destroyed at the end of the job.

vii) *Site records and minutes*
Site records and minutes should be examined before making a claim. Accurate records must be maintained on potential claims areas and these should be verified by the engineer's representatives. Keep these until after the end of the job. The accuracy and comprehensiveness of the minutes of site meetings should be checked, particularly those where the weather is wrongly blamed. Comment should be made at or before the next meeting.

Site diaries giving details of staff, plant, labour, hours worked, work done, drawings and instructions received, visitors, weather and other matters are invaluable when presenting claims.

viii) *Conversations and intentions*
If possible, conversations and intentions should be recorded. Alternatively, a record should be kept of who said what, where and when, and who was present. A duplicate book can prevent many problems but can create them if used badly.

ix) *Concealed information*
Care needs to be taken in dealing with tender documents, and claims, where alterations to any document exist. Reasons underlying the alteration should be sought and anything unusual which would indicate that information was concealed.

NOTICES GENERALLY

i) *Main notifications to the engineer.*
Table II lists the main notifications that have to be made to the engineer. Other notifications, however, may apply. The time allowed should be noted.

ii) *Engineer's reply*
Compliance with the engineer's reply is important, particularly with regard to the keeping and verification of records.

iii) *Request for instructions, drawings and details*
Think and plan ahead and ask in writing for instructions, drawings and details in good time.

iv) *Submission of programme*
The provisions of sub-clauses 14(1) and (2) should be noted, particularly the former which states;

> 'Within 21 days after the acceptance of his tender, the Contractor shall submit to the engineer for his approval a programme showing the order of procedure in which he proposes to carry out the work and thereafter shall furnish

Table II Main notifications to be made to the engineer

Clause	Brief description	Period
2 (4)	Contractor's dissatisfaction with any act of the RE or assistant or other duly authorised person.	As soon as possible
4	Request to sub-let any part of the works.	Before sub-letting
5	Ambiguities or discrepancies in the documents.	When found
7 (2)	Further information required by the contractor.	Adequate notice
11 (1)	Request for all information available.	Before tendering
12 (1)	Adverse physical conditions and artificial obstructions.	Immediately possible
13 (2)	Mode and manner of construction.	As necessary
14 (1)	Programme to be furnished together with descriptions and methods of construction.	Within 21 days of acceptance of tender
14 (2)	Revision of programme.	As soon as possible when requested
14 (3)	Method of construction.	As soon as possible when requested
15 (2)	Approval of agent.	Before commencing
21,23, (2)	Approval of insurances.	At start
26 (1) 26 (2)	Giving notices and payment of fees) Conforming with statutes, etc.)	At start and as necessary
27 (4)	Commencing work in a street, controlled land or prospectively maintainable highway. Commencing any work on the above which is likely to affect the apparatus of any owning undertaker.	Not less than 21 days beforehand
27 (7)	Other obligations of the Public Utilities Street Works Act, 1975.	As necessary
30 (3)	Damage to bridge or highway, communicating with the site, from the transport of materials, manufactured or fabricated articles.	As soon as damage known
32	Fossils.	Immediately found
35	Returns of labour and plant.	As required by the engineer
38 (1)	Examination of work before covering up.	Due notice

Clause	Brief description	Period
40 (2)	Suspension lasting more than three months.	After 3 months from the date of suspension with further notice if necessary later
44 (1)	Extension of time for completion.	Within 28 days after the cause of the delay has arisen or as soon thereafter as is reasonable in all the circumstances
45	Night and Sunday work.	Beforehand or, in an emergency, as soon as possible thereafter
48 (1)	Works substantially complete.	As soon as posible thereafter
48 (2)	Sections substantially complete or substantially part occupied or used.	- do -
51 (2)	Contractor's confirmation in writing or engineer's oral orders.	As appropriate
52 (2)	Rate or price rendered unreasonable or inapplicable.	Before work commences, or as soon thereafter as is reasonable in all the circumstances
52 (3)	Dayworks as follows: Receipts and vouchers for daywork Quotations for materials Daywork details to RE Priced statement to RE.	 As necessary Before ordering Each day End of each month
52 (4)	Notice of claims a) Contractor claiming higher rate or price than one notified under Clause 52 (1) (Valuation of ordered variations) or Clause 52 (2) (Engineer to fix rates) or Clause 56(2) (increase or decrease of rate).	Within 28 days of notification
	b) Contractor intending to claim additional payment other than under Clause 52(1) or 52(2).	As soon as possible after event, keeping records to support claim
	FOR LIST OF SOME OF THE MAIN CLAUSES FOR CLAIMING EXTRA COSTS SEE TABLE 1.	
	c) Keeping of records.	As instructed by the engineer.

Clause	Brief description	Period
52 (4)	d) Submitting of accounts.	First interim as soon as reasonable. Thereafter, as required by the engineer
53 (6)	Consent to remove plant, goods or materials.	Before removal
54 (2)	Vesting of goods or materials not on site.	As necessary
58 (6)	Production of vouchers, etc.	As requested
59A (1)	Objections to nominated sub-contractors.	As necessary
59B (2)	Termination of sub-contract.	- do -
59C	Proof of payment to nominated sub-contractors.	As requested
60 (1)	Monthly statement.	After the end of each month
60 (3)	Final account.	Not later than 3 months after date of maintenance certificate
66	Settlement of disputes.	As appropriate
69 (6)	Tax fluctuations.	As soon as possible after its event. Records to be kept. Details monthly with statement.
70	VAT.	As detailed
71	Metrication.	Immediately as appropriate

such further details and information as the engineer may reasonably require in regard thereto. The Contractor shall at the same time also provide in writing for the information of the engineer a general description of the arrangements and methods of construction which the contractor proposes to adopt for the carrying out of the works'.

All too often the contractor does not fulfil the requirements of this clause and at a later date has difficulty in substantiating reasonable extra costs.

3. Presentation of claims for entitlement

The claim must show the amount of money and extension of time claimed, and refer to the relevant contract clauses. It is best presented in three sections with a fourth section, if required, for appendices.

INTRODUCTION
The introduction should give enough information to familiarise the reader with the claim and contract details generally by giving:
- summary of claim;
- location and type of work and the main quantities;
- tender information, dates and value;
- conditions of contract and amendments;
- employer and engineer;
- specification;
- bill of quantities and sectional totals;
- fluctuations/formula.

HISTORY
The sequence of events should be given in sufficient detail to show the contractor's entitlement and responsibility for the cost and extension of time required, together with back-up information.

QUANTIFICATION
The start of the evaluation should show a summary of the amount claimed followed by the build-ups based on the history previously given, along the lines given in the preceding sections. An example is given at Appendix II.

The descriptions attached in the Conditions of Contract to the word 'cost' vary somewhat as described below and this will be noted together with the definition and interpretation given in Clause 1(5), 'include overhead costs':

> Reasonable cost with reasonable additions —
> Reasonable and proper — fair and reasonable

The wording of the contractor's entitlement to costs in the Conditions of Contract varies somewhat in the description. Nevertheless, it is consistent that they shall be reasonable even to the words 'reasonable percentage addition thereto in respect of profit' (Clause 12(3)).

The engineer has the power to certify for reasonable costs in a number of clauses.

TIME FOR PRESENTATION
There are two main types of presentation:

i) *Final claim*
The final claim can be made where sufficient details are available to finalise, or nearly finalise, the repercussions of the incident, leaving the engineer to settle either then or later, or part then and part later.

ii) *Interim claim for payment on account*
Often, although factors are not known, some payment on account is needed. In this case an interim claim is submitted. This gives the engineer an opportunity to start his investigations and should expedite settlement of the final claim when the full extent of the delay and reasonable extra cost may be decided.

INCORPORATION IN CERTIFICATES: INTEREST: INCREASED COSTS
The claim, when submitted, should be included in the next monthly statement and when payments are made, interest on overdue payments (Clause 60(6)) claimed.

Where the contract includes a formula method of dealing with increased costs, such as Baxter, the amount cannot be moved above the Baxter index figure until certified. Prompt investigation and certification by the engineer thus reduces the amount of adjustments required later as well as helping the contractor's cash flow.

The practice of delaying payment on claims as long as possible, particularly to force a contractor into a cheap settlement, or to force him out of business, is to be deprecated.

CONCILIATION
With most claims agreement is soon reached between the parties, particularly when the engineer acts absolutely impartially, without fear or worry of the client disputing his decision later.

In other cases the claim falls into a 'grey' area. Often, conciliation between the parties may be effected by a third party representing the contractor's case to the engineer, without resort to the cost of arbitration.

ARBITRATION
When it seems that both parties are not prepared to move from positions, a Clause 66 decision must be obtained and consideration be given to arbitration.

Care must be taken to observe the actual wording of the Clause 66 for the contract concerned. The wording in the January 1986 reprint is substantially different from the previous revision. The new wording was introduced by Guidance Note 6 (June 85) of the Conditions of Contract Standing Joint Committee. This note gave approved amendments to Clauses 66 and 67 and recommended that these amendments be included in all contracts entered into on or after 10 June 1985.

It was recommended that Clause 66 be amended to state that arbitration should be conducted in accordance with the Institution of Civil Engineers' Arbitration Procedure (1983).

The following comments were also made in that Guidance Note 6 (Arbitration) on the Clauses 66 and 67 recommended in that note.

Dealing first with the proposed Clause 66, the note stated:

> 'If a dispute or difference arises between the Employer and the Contractor and a Certificate of Completion of the whole of the Works has not been issued the Engineer's Clause 66 decision is to be given in writing within one month of the request for such a decision. If a dispute or difference arises after the issue of a Certificate of Completion of the whole of the Works the Engineer's decision is to be given within three months of the request for such a decision.
>
> The Clause also provides that unless the Parties otherwise agree in writing, any dispute or difference which may arise during the course of the Works may be referred to arbitration notwithstanding that the Works are not then complete or alleged to be complete.'

and went on to state:

> 'Clause 67 has been amended to state that if the Works are situated in Scotland any reference to arbitration should be conducted in accordance with the Institution of Civil Engineers' Arbitration Procedure (Scotland) 1983.'

The January 1986 reprint incorporated the recommended clauses 66 and 67 and sub titles were added in the margin.

Contracts entered into after 10 June 1985 may contain either the old or the new clauses and this requires to be noted.

When the old Arbitration Procedure of 1973 was replaced in 1983, it was by two new procedures; one for England and Wales and one for Scotland. The aim of the new procedures was to speed up construction arbitrations and possibly reduce the costs by giving more power to the arbitrator, introducing summary awards and three special procedures for:

a) interim arbitrations;
b) experts; and
c) a short procedure.

It may be advantageous in arbitrations under the old Clause 66 to have this 1983 procedure used either by agreement between the Parties or by the President so directing or under the conditions given in the Procedure Rule for the Arbitrator to so stipulate provided the Parties do not agree otherwise.

Following the new 1983 procedure came a revised Form Arb ICE for use in selecting the arbitrator to be appointed. The form has five parts:

Part 1 Details of the Contract(s);
Part 2 Notice to Refer Dispute(s) or Difference(s) to Arbitration;
Part 3 Notice to Concur in the Appointment of an Arbitrator;
Part 4 Application to the President ICE to appoint an Arbitrator;
Part 5 Appointment of an Arbitrator by the President ICE.

The form is applicable to ICE and FIDIC Conditions of Contract, the FCEC Form of Sub-Contract and may be adapted for other contracts. Under 'Notice to Concur' only one person's name has now to be put forward for consideration by the other party as the arbitrator. It should be remembered that notice is served on the employer and not on the engineer (vide Clause 68(2) re the Service of Notice on the Employer).

The powers given to the arbitrator in the 1983 procedure include power to order the deposit of money or security, power to debar a party in default and power to proceed with the hearing in one party's absence.

The Summary Awards are analogous to Order 14 of the Rules of the Supreme Court with, hopefully, early redeployment of money. The arbitrator acts on appropriate admissible evidence, may direct affidavits in support and in opposition be served in advance of the date for hearing the application. Main use where there is a sum owing as to which there is no defence (even where no specific sum may be identified as owing). Liability must not be in issue or a set off to extinguish the claim.

The short procedure is for minor and straightforward disputes and basically is on documents with a site visit and oral submissions or questions. Normally no costs are awarded; there is no right of cross examination and there is little opportunity for counsel. One party may, in the circumstances given, cancel the use of the short procedure but with a liability on costs. On small claims this procedure has the defect that the winner has to carry his own costs. He may lose on small claims due to this.

The Expert's Procedure is not dissimilar to the short Procedure, but it is for experts and for technical issues. It is based on documents, a site visit and oral submissions or questions. There is no provision for going back as in the short Procedure, and costs, but not normally legal costs, may be awarded. It is really a formal meeting between the arbitrator and experts but without the formal trappings of a court room. The hearing is without professional advocates but they may advise their experts on questions to ask.

The Interim Arbitration Procedure is to enable an arbitrator to make decisions before completion. A speedy hearing is essential. The aim is generally to give finality, and to affect the engineer's power as little as possible. ICE Clause 63 (forfeiture) disputes are excluded. The site is seen at the time.

Provision is also made for referring further disputes to the arbitrator, and for the arbitrator to use his own technical knowledge. The old 1973 procedure dealt, mainly, with the forms for appointing an arbitrator and correspondence from the arbitrator. These items are now dealt with by a manual of guidance[12]. The 1973 procedure book recorded that in Scotland 'the arbiter's powers are derived from Common Law' and 'Part I of the Arbitration Act 1950 does not apply'.

Some authorities convey the impression that the costs of the hearing will be excessive and thereby influence the contractor to accept an unreasonable offer. In this case the contractor has to rely on the arbitrator's ability to control the hearing and to make an appropriate award on costs.

At present it is too easy to cover up someone's wrong decisions by having an unsuccessful arbitration and blaming the arbitrator for his decision. Auditors might well examine this matter and the cost.

JUDGEMENT PROBLEMS

Some authorities delete the arbitration clauses and a contractor should consider whether it is wise to tender on such a document. On civil engineering contracts, generally, an arbitrator has the power to open up, review and revise any decision, opinion, instruction, direction, certificate or valuation of the engineer. If the arbitration clauses are deleted this power is not there. Further the courts have to be used.

Even when there are arbitration clauses in the contract, it is questionable whether the courts have that power of review and revision referred to above. Without these clauses there is little doubt that they have not that power.

The observation of the Court of Appeal in the *Northern Regional Health Authority* v *Derek Crouch Construction Co Ltd* (1984) 1QB644 was that a court did not have those powers of review and revision for the contract considered (JCT). As liability is more important than quantum, the courts are severely handicapped at present on a main contractor's claim.

The court's position may be changed as a result of a decision in *Partington & Son (Builders) Ltd* v *Tameside MBC* delivered on 4 November 1985, where it was held that the power of an arbitrator under the arbitration clause is no different from that possessed by a judge if the court dealt with the dispute. This should not be relied upon as yet in view of the observation given above by the Court of Appeal.

STAYING COURT PROCEEDINGS

The Arbitration Act 1950, where there is a written arbitration clause, gives power for the staying of court proceedings as follows:

> '*Staying court proceedings where there is submission to arbitration*
> 4. (1) If any party to an arbitration agreement, or any person claiming through or under him, commences any legal proceedings in any court against any other party to the agreement, or any person claiming through or under him, in respect of any matter agreed to be referred, any party to those legal proceedings may at any time after appearance, and before delivering any pleadings or taking any other steps in the proceedings, apply to that court to stay the proceedings, and that court or a judge thereof, if satisfied that there is no sufficient reason why the matter should not be referred in accordance with the agreement, and that the applicant was, at the time when the proceedings were commenced, and still remains, ready and willing to do all things necessary to the proper

conduct of the arbitration, may make an order staying the proceedings.'

The words 'before delivering any pleadings or taking any other steps ..... ' should be particularly noted. Thought should be given to this before issuing a writ for an Order 14 Summary award under the Rules of the Supreme Court. It might prevent a stay for arbitration.

The powers listed above given to an arbitrator by the arbitration clauses are essential for the satisfactory settlement of most disputes. Main contractors require to be eternally vigilant not to lose these powers from the case.

LAW, PLEADINGS AND DISCOVERY
The law used in an arbitration is normally clear from the provisions of the contract and the same may apply on procedure. If this is not so, the problem is outside the terms of this publication. If agreement cannot be reached between the parties to the contract the Courts may have to decide.

Pleadings may take some time and a good arbitrator will attempt to limit the delay. Pleadings consist basically of the following, plus requests for further and better particulars:

points of claim;
points of defence (and counterclaim if any);
points of reply (and defence to counterclaim if any);
points of reply to defence to counterclaim if any.

Discovery (ie, disclosure) and inspection of documents follows. This emphasises the care that should be taken to ensure that site documents are not destroyed when the site is cleared. Expert witness reports are then normally exchanged. Frequently, the statements of the witnesses as to fact are also exchanged before the hearing, with the provisions of the rules of the Supreme Court Order 38 Rule 2A applying.

Details of arbitration are given in a number of textbooks[7,8,11] in Arbitration Acts[9] (where applicable) and in a number of publications of the Chartered Institute of Arbitrators.

FURTHER OPPORTUNITY FOR SETTLEMENT
Settlement between the parties is frequently reached during the pleadings stage when the respective cases are studied perhaps in greater detail. The arbitrator will frequently assist in this.

DELAYS IN PAYMENT
Of considerable concern to most contractors is the cash flow position. If the cash is not available they are not able to take on new work. Money employed will be turned over several times a year and must produce more than the rate of interest allowed in accordance with Clause 60(6) to keep the firm going.

Denying money to the contractor is both unfair and immoral. The fact that the cost of borrowing money is often greater than the payment in accordance with Clause 60(6) is no excuse.

Delays occur:

a) with the engineer who sometimes wishes to see how things are going financially before making a decision or with subordinates who withhold agreement in an effort to force a cheap settlement;

b) with the employer who often has committees causing delays, apart from his own cashflow problems;

c) with auditors who take an unreasonable time to examine the accounts. Where they are overworked a payment on account should be made, as frequently a large sum is held up for an unreasonable time. This is particularly so where they are disputing an engineer's decision, without asking the employer to obtain a Clause 66 decision. The contractor may eventually have to do this, thereby causing further delay. The auditors are not a party to the contract, which causes further problems;

d) with third party authorities not named in the contract, who employ agent authorities named as the 'Employer' in the contract. The latter should make the payment but, unfortunately, this is not always the case. Payment by the third party authority is often delayed for reasons which have nothing to do with the contractor. It may be an argument between the authorities as to responsibility, or the auditors of one authority querying the decision of the 'Engineer' employed by the other authority. Legal action takes time, particularly with the delays which may be involved when the employer does not want, or may not be able, to part with money he has not recouped from the third party authority;

e) with arbitration which inevitably takes some time.

Contractors have to increase their overhead additions to allow for the above which is unfair to the prompt payer. In effect, this means that one client will subsidise others.

INTEREST

Where a claim for entitlement is settled without recourse to arbitration or the courts, interest may be payable as a 'reasonable cost' to the contractor or under Clause 60(6), as appropriate.

Where arbitration takes place the arbitrator may allow for the interest up to the date of his award. Where the Arbitration Act 1950 applies the arbitrator has power to award interest under Section 19A, inserted in that Act by the Administration of Justice Act 1982, as follows:

> 'Power of arbiter to award interest
>
> 19A - (1) Unless a contrary intention is expressed therein, every arbitration agreement shall, where such a provision is applicable to the reference, be deemed to contain a provision that the arbitrator or umpire may, if he thinks fit, award simple interest at such a rate as he thinks fit —
>
> a) on any sum which is the subject of the reference but which is paid before the award, for such period ending not later than the date of the payment as he thinks fit; and
>
> b) on any sum which he awards, for such period ending not later than the date of the award as he thinks fit.
>
> (2) The power to award interest conferred on an arbitrator or umpire by subsection (1) above is without prejudice to any other power of an arbitrator or umpire to aware interest.'

In addition Section 20 states:

> 'A sum directed to be paid by an award shall, unless the award otherwise directs, carry interest as from the date of the award and at the same rate as a judgement debt'.

COST TO THE CONTRACTOR OF CLAIMS FOR ENTITLEMENT

A contractor often has to allow in his overheads for the cost of producing claims which are not settled by arbitration or the courts.

Where settled by these courts, the award will normally give directions on how the costs are to be apportioned between the parties. Some employers pay heavily for the failure of those more concerned with keeping a contract within its budget price than with the contractor's contractual entitlement.

A contractor has to assess the cost of a claim for entitlement against the money he will probably obtain. With small amounts, star rates and star items may be negotiated at little cost. With some authorities he may tender a higher price for the same work, particularly with those known to be unreasonable.

There is no easy answer. The contractor will have to consider each claim on its merits and review the claim as it progresses. A badly presented claim, not in accordance with the points discussed earlier, is money lost from the start.

COUNTERCLAIM BY THE EMPLOYER

It is not unusual for contractors to receive a counterclaim from the employer, either in the early stages of his claim or at the pleadings stage.

Sometimes, a contractor receiving such a claim may consider that the employer's representatives drew it up 'tongue in cheek' and tends to discount the counterclaim.

He must, however, treat the counterclaim very seriously, refuting and dealing with it accordingly. If he does not, he may find that part of it will 'stick' even when there is no justification.

4. Contract clauses and claims for entitlement

Many of the clauses of the ICE contract have a direct or indirect effect on a contractor's entitlement to extension of time or reimbursement of costs resulting from the 'employer's responsibility'. The following clauses have been selected for comment. Such comment is supplementary to the text of the Conditions of Contract, the aim being to amplify what is contained therein.

Clause 1(1) — Definitions
The importance of Clause 1(1) is that it states who is the 'Employer' and who is the 'Engineer'. It should remove doubt. The 'Employer' may be an agent authority for another organisation which, in practice, may do the paying or whose auditors may attempt to intrude on the contract. That organisation and its auditors are not, nevertheless, parties to the contract between the 'Employer' and the 'Contractor'. They must not be allowed to intrude or influence the 'Engineer' in holding the balance between the parties and this must never be forgotten. The 'Engineer' is a specific person and not one of his subordinates or a conglomerate. The responsibility for decision is his. The other definitions are useful.

Clause 1(5) — Cost
Cost is here deemed to include overhead costs, whether on or off the site, except where the contrary is expressly stated.

Clause 2 — Engineer's representative
Clause 2 (1) defines the somewhat limited power of the engineer's representative and Clause 2(1) gives power to appoint assistants.

Clause 2(3) enables the engineer to delegate certain of his powers to the engineer's representative but it should be noted that prior notice in writing of any such authorisation shall be given by the engineer to the contractor.

Clause 2(4) lays down the procedure for a contractor who is dissatisfied by reason of an instruction or an act of these people.

Clause 4 — Sub-letting
Written consent of the engineer must be obtained before the contractor sub-lets any part of the works.

Clause 5 — Documents mutually explanatory
This is an important clause from the claims aspect since it states that ambiguities or discrepancies shall be explained and adjusted by the engineer in writing, which shall be regarded as instructions issued in accordance with Clause 13. Clause 13(3) can provide the basis for a monetary settlement as well as time.

Clause 7(1) — Further drawings and instructions
The engineer is authorised to supply modified or further drawings and instructions. It is important to keep a record of these drawings as they arrive, showing the date of arrival and the main differences. There will probably be a claim for entitlement on the changes and also for late arrival.

Clause 7(2) — Notice by contractor
It is vital for the contractor to review frequently the information needed and give adequate notice in writing to the engineer of any drawings or specifications required. The notice must be reasonable. It is unwise not to give this notice, since if the engineer is late in supplying he will simply say the contractor never asked for them. Many a good claim has been lost this way.

Clause 7(3) — Delay in issue
This deals with delay in the issue of drawings or instructions requested by the contractor but note the words 'and considered necessary by the Engineer'. This may cause difficulty and argument. Also note the words, 'If such drawings or instructions require any variation to any part of the Works the same shall be deemed to have been issued pursuant to Clause 51' (Clause 51 deals with ordered variations). Keep a record of the dates the drawings and instructions arrive.

Clause 8(1) — Contractor's general responsibilities
Under the contractor's general responsibilities is the requirement for providing anything 'specified in or reasonably to be inferred from the Contract'.

Clause 8(2) — Design
The contractor is not responsible for design or specification of the permanent works except as provided and of temporary works designed by the engineer.

Clause 11(1) — Inspection of site
Earlier in this paper, emphasis was given to the importance, at the tender stage, of examining all information recorded as being available. Those compiling tenders must comply fully with this clause. After the words 'nature of the ground and sub-soil', note the words in brackets '(so far as is practicable ... Employer)'.

Clause 12(1) — Adverse physical conditions and artificial obstructions
It should be noted whether this clause, dealing with adverse physical conditions and artificial obstructions, has been altered or mutilated. If so, find out exactly what the change is, consider why it has been made, and what the repercussions may be. It may not be worth the risk of tendering for the job, or a higher price may be necessary.

This is a very useful clause when things turn out to be different from that anticipated but the event must be covered by the wording. It is important to 'give notice to the Engineer pursuant to Clause 52(4)' and specify as stated in this Clause 12(1). Be brief but cover the essentials with further information later when determined. Note in Clause 52(4) the necessity to keep records. Do not forget to claim the cost and feed the amount into the monthly statement. If there is delay in settling you have staked your entitlement to interest.

Clause 12(2)(d) — Suspension under Clause 40
Ensure that any suspension under Clause 40 is given in writing.

Clause 12(3) — Delay and extra cost
Note that the contractor is entitled to the reasonable cost of any additional constructional plant used, and that the entitlement is given for a reasonable percentage addition in respect of profit and for 'the reasonable costs incurred by the Contractor by reason of any unavoidable delay or disruption of working suffered as a consequence of encountering the said conditions or obstructions for such part thereof'.

Clause 12(4) — Variations previously ordered
This clause gives the contractor entitlement to payment for any variations previously ordered under Clause 12(2)(d).

Clause 13(1) — Instructions and directions
Note the words 'The contractor shall take instructions and directions only from the Engineer or (subject to the limitations referred to in Clause 2) from the Engineer's representative'.

Clause 13(2) — Mode and manner of construction
The words 'conducted in a manner approved of by the Engineer' are important and should be read in conjunction with Clause 14.

Clause 13(3) — Delay and extra cost
Note the last sentence and that the engineer, in theory, should take action on this sub-clause on delay and extra cost himself without notification from the contractor other than later in the monthly statement (Clause 60(1)).

Clause 14(1) — Programme to be furnished
The submission of a programme in accordance with this clause is important. Note also the second half of this sub-clause dealing with a general description.

Clause 14(2) — Revision of programme
If delay, which is the responsibility of the employer, has occurred and the engineer wishes to keep the same completion date, then a claims situation exists.

Clause 14(4) — Engineer's consent
The words 'reasonable period' depend on the circumstances but a claims situation may be produced by delay.

Clause 14(6) — Delay and extra cost
Delay and extra cost resulting from Clause 14(4) and (5) is covered. The engineer should take action on this himself without prompting by the contractor, except later in so far as Clause 60 is concerned.

Clause 15 — Necessary superintendence
Necessary superintendence must be provided but payment will depend on responsibility for delay, disorganisation or disruption.

Clause 17 — Setting-out
The setting-out requirements may be increased. If this is the employer's responsibility the extra reasonable cost is reclaimable.

Clause 19(1) — Safety and security
Additional safety and security, where they are the employer's responsibility, is reclaimable.

Clause 19(2) — Employer's responsibilities
A claims situation may arise here.

Clause 20(2) — Responsibility for reinstatement
Note the words 'To the extent that any such damage loss or injury arises from any of the excepted Risks the Contractor shall if required by

the engineer repair and make good the same as aforesaid at the expense of the Employer'. The 'Excepted Risks' are defined in Clause 20(3) and include 'a cause due to use or occupation by the Employer, his agents, servants or other contractors (not being employed by the Contractor) of any part of the Permanent Works or to fault, defect, error or omission in the design of the Works (othere than a design provided by the Contractor pursuant to his obligations under the Contract)'.

Where other people cause damage, loss or injury to the works or any part thereof while the contractor is responsible, he should notify them of the damage by recorded delivery and bill them for it. This applies particularly to service and other undertakers not covered by the excepted risks. Too frequently the contractor is left paying for their damage to the works.

Clause 21 — Insurance of works, etc.
Insurance of the works may cost more due to the 'responsibility of the Employer', but insurances are normally included in the overheads which are adjusted. The same applies to Clauses 23 and 24.

Clause 22 — Damage to persons and property
There is sometimes a certain amount of disagreement on where Clause 22(1)(b)(iv) ceases to cover the contractor, particularly where adverse physical conditions and artificial obstructions covered by Clause 12 are involved. With a reasonable engineer there should be no difficulty — however, problems are often experienced with his subordinates.

Clause 26(1) — Giving of notices and payment of fees
This sub-clause enables the contractor to be reimbursed for the rates on his site huts, as well as a number of other items.

Clause 26(2) — Planning permission
This is a useful indemnity on planning permission.

Clause 27 — Public Utilities Street Works Act 1950 — Delays
Delay may result from the effects of the Public Utilities Street Works Act 1950 and an extension of time may be awarded. The cost may be reimbursible in accordance with sub-clause (6), in theory, without a request from the contractor other than on the monthly statement (Clause 60).

Clause 30(3) — Damage to bridges and highways
This clause lays down the rules for payment for damage to bridges and highways. Alterations to the works may affect the amounts involved and a claims situation may arise.

Clause 31 — Facilities for other contractors
Delay and extra cost caused by the 'Other Contractors' referred to in sub-clause 31(1) are covered by sub-clause 31(2). Some service undertakings have caused very serious delays in the past. It is as well, in the early stages of a delay, to ensure that Clause 31(1) covers the contractor over the delay in question. The intention of Clause 31(1) is clear and the wording good and fairly inclusive. One cannot, however, be absolutely sure of some of the fringe cases without checking with the engineer when time is still on one's side. In theory, no notification by the contractor is necessary, other than in accordance with Clause 60 later on the monthly statement. In practice, notification is well worthwhile so that there is no dispute later on the facts.

Clause 32 — Fossils, etc.
Fossils, etc, sometimes cause considerable delay to contracts, particularly where suspension of part of the works under Clause 40 is involved. Be sure the suspension order is in writing. Reasonable cover to the contractor is given by the words at the end of that clause.

Clause 33 — Clearance of site on completion
Extra reasonable cost is reimbursible under this clause when caused by the employer's responsibility.

Clause 35 — Return of labour and plant
It is up to the engineer to state the returns of labour and plant he requires. These are often used in support or otherwise of claims for entitlement later.

Clause 36 — Samples and tests
This clause lays down when costs of samples and tests may be reclaimed by the contractor.

Clause 38 — Examination of work before covering up
Sub-clause 38(1) deals with the contractor giving due notice to the engineer before covering up work. It never pays to be lax over this sub-clause even if the named engineer is normally miles away and could not comply, and the engineer's representative has not been authorised to act in accordance with Clause 2, or may even be only part-time. It is dangerous to take anything for granted. The engineer might claim later than he would have flown in. In any case, the contractor may later wish to claim under (the last part of) Clause 38(2).

Clause 39 — Removal of work and materials
Clause 39 gives the engineer wide powers over the removal of work and materials and this may be the subject of dispute later. It should be remembered that Clause 1(3) states, 'The headings and marginal notes in the Conditions of Contract shall not be deemed to be part thereof or be taken into consideration in the interpretation or construction thereof or of the Contract'. The word 'improper' only appears in the marginal note and one has to fall back on intentions and inferred terms. Fortunately, the engineer's attitude is generally reasonable over this.

Clause 40(1) — Suspension of work
Suspension of the works, as opposed to part of the works, is rarely ordered. It is essential to ensure that any order is given in writing. It is more often used as a threat by the engineer. The clause is clear on cost and delay.

Clause 40(2) — Suspension lasting more than three months
Optional action open to the contractor, when the suspension lasts more than three months, is given. The effects of omission are dealt with in Clause 51. Action on abandonment depends much on the circumstances and reasons. It may be settled amicably at one extreme or go to breach of contract proceedings at the other. The problem is not met frequently and is outside the terms of this paper particularly as other factors such as the state of the employer's finances may be involved. Negotiations often start before the three months is up.

Clause 41 — Commencement of the works
The period after the acceptance of the tender to the commencement of the works may be long, particularly where land acquisition is concerned. This clause gives the contractor the opportunity to re-negotiate an extra for the delay. Conditional acceptance awaiting say Ministry, Council or other approval is not acceptance and it may be necessary to commence negotiations if full acceptance is delayed too long after the submission of the tender before the acceptance goes through — or, to give notice of the contractor's wish to do so.

Clause 42 — Possession of the site
Delay in obtaining possession of the site and particularly part of the site is not unusual. The delay and reasonable cost to be awarded to the contractor are dealt with in this clause. The contractor, in theory, does not need to make any claim other than in accordance with Clause 60 in the monthly statement. In practice, he is well advised to do so, as the repercussions of these delays and the disorganisation, disruption and out of sequence working resulting are often very considerable. Only the contractor knows their effect fully. The importance of furnishing the programme in accordance with Clause 14 is clear.

Clause 44(1) — Extension of time for completion
To avoid difficulties on this clause, a contractor should always give notice of every delay to the engineer within 28 days after the cause of the delay has arisen. Where not given within that time, then as soon thereafter as is reasonable in all the circumstances in accordance with this sub-clause.

Many clauses in the contract lay down that the engineer shall take 'such delay' into account in determining any extension of time, to which the contractor is entitled under Clause 44, and it would appear that the engineer should take action himself.

The opening wording to Clause 44, after referring specifically to certain sub-clauses, is nevertheless so all-embracing that it would be unwise not to notify every delay. Subsequent sub-clauses of this clause are, however, quite clear in that the engineer has the power to assess an extension of time without a claim from the contractor.

Deleting work from a programmed contract rarely saves time. A similar amount of work added elsewhere in a contract normally delays.

It will not normally be possible to give the full repercussions of each delay on the job in the early stages. The repercussions of some delays are not great, but nevertheless the claim should be staked.

Clause 44(2) — Interim assessment of extension
This sub-clause places on the engineer a responsibility, during the running of the contract, to assess from time to time the extension of time (if any) to which he considers the contractor is entitled, whether he has had a claim for extension of time or not.

The contractor must ensure that he is not given an extension of time for adverse weather when the cause is really the responsibility of the employer, with a cost element reimbursible to the contractor. Note that the engineer must inform the contractor where he does not consider the contractor is entitled to an extension of time on a claim.

Clause 44(3) — Assessment at due date for completion

This is similar to clause 44(2) but is for the assessment at due or extended date of completion. Where the engineer considers that the 'Contractor is not entitled to an extension of time he shall so notify the Employer and the Contractor'.

Clause 44(4) — Final determination of extension

This deals with the final determination of the extension. Note that 'No such final review of the circumstances shall result in a decrease in any extension of time already granted by the Engineer pursuant to sub-clauses (2) or (3) of this Clause'. This latter wording, unfortunately, makes some engineers drag their feet to the contractor's detriment.

For this reason the contractor should continue to make his claim for the full amount in the monthly statement submitted in accordance with Clause 60(1) so that at least he may obtain interest on the money overdue when paid. It is unfortunate that fear of being wrong, and this being commented upon by the auditors later has this effect. Nevertheless, the inclusion of the wording is better than nothing as many engineers face up to their responsibility reasonably fearlessly and with great integrity. The problem is greater when engineers to large authorities become too dependent on their subordinates and do not always have time to consider such details in sufficient depth until the claim reaches a Clause 66 state.

Clause 45 — Night and Sunday work

A problem arises here when an engineer, in response to a request to work nights or Sundays, replies neither saying yes or no but referring the contractor to some law of the country such as one on noise or other environmental problem. Such replies are extremely unsatisfactory. If the contractor claims later for not being allowed to work he is told permission to work has not been refused and if he starts to work he is told he has not had permission. The contractor is advised to write back for a definite decision subject to his complying with the law. Unfortunatley, by then delay has occurred.

This practice of negative control of contracts by an engineer may also occur with other clauses. It may lead the contractor into serious losses which he may then find he is unable to reclaim. He must constantly be on his guard when he comes across any engineer employing negative control.

Clause 46 — Rate of progress

It will be noted that where the engineer considers the rate of progress to be too slow, permission for night and Sunday work 'shall not be unreasonably refused'. The first line of this Clause should also be noted.

Clause 47(4) — Deduction of liquidated damages

Liquidated damages may not be deducted until action has been taken by the engineer in accordance with Clause 44(3) or (4). This means until after the due date or extended date for completion of the works or relevant section. The engineer must be 'of the opinion that the Contractor is not entitled to any further extension of time'.

Clause 48 — Certificate of completion of the works

The onus is on the contractor to ask for the certificate of completion of the works in writing for action by the engineer within 21 days of the date of delivery. This clause may be particularly useful on contracts which, towards the end, never seem to reach finalisation. Completion of sections, occupied parts and 'other parts' of the works are also dealt with.

Clause 49 — Work requiring repair, etc.

Work requiring repair, etc, must be inspected before the expiration of the period of maintenance and be required of the contractor within 14 days of its expiration. Sub-clause (3) lays down when the contractor shall be paid. Sub-clause (5) deals with reinstatement.

Clause 50 — Contractor to search

This clause indicates when the contractor shall be paid for searches.

Clause 51 — Ordered variations

This clause, dealing with ordered variations, is most important from the claims for entitlement aspect. Orders must be in writing or, if oral, confirmed in writing by the engineer, or by the contractor and not contradicted in writing by the engineer. In sub-clause 51(1) note the words 'for the satisfactory completion and functioning of the works'. The engineer may not vary the character of the works as a whole under this clause. It is somewhat restricting on the engineer and the contractor must guard his rights. Remember the effect on time and cost of variation orders is often considerable and may not be apparent at the time. Constant monitoring of progress is necessary.

Clause 52(1) — Valuation of ordered variations

This is a most important clause. Note in the first

sentence the words 'after consultation with the contractor' and in the second sentence the words 'of similar character and executed under similar conditions'. Note also the words 'fair valuation'.

Clause 52(2) — Engineer to fix rates
This clause covers cases where variations render rates unreasonable or inapplicable. Notice is to be given 'before the varied work is commenced or as soon thereafter as is reasonable in all the circumstances'.

Clause 52(3) — Daywork
Note that with variations on daywork the contractor 'before ordering materials shall submit to the engineer quotations for the same for his approval'. Note also the list to be delivered daily to the engineer's representative and the monthly priced statement with the alternative available to the engineer.

Clause 52(4) — Notice of claims
This clause deals with the Notice of Claims and is disregarded at the contractor's peril. Sub-clause (a), dealing with the contractor's requirement for higher rates or prices than notified, states 'within 28 days', 'in writing' and his 'intention' and means precisely that.

Sub-clause (b) states:

> 'If the contractor intends to claim any additional payment pursuant to any Clause of these Conditions other than sub-clauses (1) and (2) of this Clause he shall give notice in writing of his intention to the Engineer as soon as reasonably possible after the happening of the events giving rise to the claim. Upon the happening of such events the Contractor shall keep such contemporary records as may reasonably be necessary to support any claim he may subsequently wish to make'.

The golden rule is, if in doubt, submit a claim for entitlement. Many clauses indicate that no claim is necessary and the engineer will act himself. In practice, he may not realise that the contractor has entitlement. Further, sub-clause (b) is wide in its description and says 'any clause'.

Note the necessity to keep records.

The amount claimed, referred to in sub-clause (d), should be fed into the next monthly statement submitted in accordance with Clause 60(1) and, in due course, interest obtained.

Sub-clause (e) entitles the engineer to certify some payment and (f) to certify 'on account'.

Clause 55 — Correction of errors
Sub-clause (2) deals with the correction of errors or omissions in the bill of quantities by the engineer.

Clause 56 — Increase or decrease of rate
Sub-section (2) is most important since it deals with entitlement to variation in rate when the quantity varies from that billed. There are many cases where an increase in the quantity may involve an increase in the rate. Examples are work involving provision of materials where either a limited amount of material only was available at a low rate, or where the obtaining of more material in a limited time involves more suppliers at higher rates. Shuttering may be used less efficiently with more shuttering used fewer times and so on. This sub-clause enables the engineer, after consultation with the contractor, to determine an appropriate increase or decrease of any rates or prices rendered unreasonable or inapplicable in consequence thereof.

Clause 57 — Method of measurement
This clause states the method of measurement to be used. If a different method of measurement is to be used, the standard clause requires amendment.

Different rates are normally inserted against similar descriptions with different methods of measurement. The method stated must be noted, as well as any variations from normal which are listed in the bill of quantities and the specification.

Both the January 1979 contract revision and the 1986 reprint nominate the *Civil Engineering Standard Method of Measurement* approved by the ICE and the FCEC in association with the ACE in 1976 or such later or amended edition thereof as may be stated in the Appendix to the Form of Tender. The edition used must be noted in that Appendix.

The First Edition CESMM (1976) had a useful guide to this then new method of measurement under the title *Measurement in contract control* published by the Institution of Civil Engineers[13].

The second edition 1985 is not meant to be a radical departure from the first edition, but to be an update and general overhaul. The foreword to the edition explains the noticeable changes. Sewer renovation and ancillary works appears as Work Classification Class Y, and Notes are now called rules. The rules have been divided into four categories by functions and are so indicated. The guide to this edition is known as the *CESMM2 Handbook*[14].

Clauses 58/59A/59B/59C — Provisional and prime cost sums and nominated sub-contracts
The difference between the provisional sum and a PC item is to be noted from the definitions at the start of Clause 58. Problems arising are governed, but not necessarily resolved, by clauses 58/59A/59B and 59C. Sub-clause 58(3) deals with design requirements.

The contractor must not be afraid to object to a nomination.

Loss, expense or damage arising from 'Direction by the Engineer' is dealt with in sub-clause 59A(3)(b).

Forfeiture and termination of sub-contracts, together with delay, extra cost and recovery, are fully detailed. Termination without consent may prove costly (Clause 59B).

Clause 60 — Certificates and payment
In the monthly statement all claims for entitlement should be included under clause 60(1)(d) so that interest may be subsequently claimed when payment is made.

Interest on overdue payments is covered by sub-clause 60(6) which in the January 86 reprint states:

> 'In the event of failure by the Engineer to certify or the Employer to make payment in accordance with sub-clauses (2) (3) and (5) of this Clause the Employer shall pay to the Contractor interest upon any payment overdue thereunder at a rate per annum equivalent to 2% plus the minimum rate at which the Bank of England will lend to a discount house having access to the Discount Office of the Bank current on the date upon which such payment first becomes overdue. In the event of any variation in the said Minimum Lending Rate being announced whilst such payment remains overdue the interest payable to the Contractor for the period that such payment remains overdue shall be correspondingly varied from the date of each such variation.'

A lower rate of interest was shown in the January 1979 revision. The ICE Conditions of Contract Joint Standing Committee in their Guidance Note No 4[15] made amendment to the then clause 60(6) and recommended that the amendment, which included the above higher rate, should appear in all contracts entered into after 1st May 1982.

It is, therefore, essential to check the wording of this clause in the actual contract concerned and the rate of interest.

Under sub-clause 60(2), the last date for payment is 28 days after the date of delivery of the contractor's monthly statement. Under sub-clause (3), the last date for payment is 3 months after receipt of the final account and of all information reasonably required for its verification plus 28 days or, if certified before the 3 months, 28 days after the date of the certificate.

Under sub-clause (5)(a) and (b) the last date for payment is 14 days after the appropriate certificate and under sub-clause 5(c) 14 days after the expiration of the period of maintenance, less an amount for outstanding work. Payment first becomes overdue the next day.

Some engineers attempt to avoid the interest payment requirement by the use of technicalities which are to be deplored. Because of this, entitlement is not always fully obtained and the clause is often failing in its purpose.

Sub-clause 60(7) deals with correction and withholding of certificates and states:

> 'The Engineer shall have power to omit from any certificate the value of any work done goods or materials supplied or services rendered with which he may for the time being be dissatisfied and for that purpose or for any other reason which to him may seem proper may by any certificate delete correct or modify any sum previously certified by him'

to which there are two provisos affecting nominated sub-contract work.

Clause 61 — Maintenance certificate/unfulfilled obligations
Sub-clause 61(1) deals with the issue of the maintenance certificate and sub-clause 61(2) with unfulfilled obligations.

Clause 63 — Forfeiture
Only as a last resort and after the most expert and experienced advice should a contractor abandon a contract. It is normally an unwise action to take except where so provided, eg, Clause 40(2).

Clause 65 — War clause
This is not considered here but gives very wide coverage.

Clause 66 — Settlement of disputes — arbitration
A dispute or difference must first arise before the

implementation of this Clause. When a Clause 66 decision is required it must be clearly stated and that clause fully compiled with, both before and after the decision and particularly as regards timings.

When adjudicating under this Clause, the engineer does not act as agent of the employer but uses his personal judgement to apply the contract fairly to the dispute or difference. He does not now act as a quasi arbitrator and this reduces his legal protection. Sometimes, he views the problem differently from previously and gives a different answer.

Often 'non-Clause 66' engineer's decisions are made by subordinates for whom the engineer is responsible, but without his actual knowledge. This is particularly so in large authorities.

The engineer may wish to give a further 'non-Clause 66' decision himself before he is asked for a Clause 66 decision. This has the advantage that it is the engineer's non-Clause 66 decision and not that of a subordinate. This is particularly advantageous when there has been a clash of personalities between the subordinate and one or more of the contractor's staff, provided the engineer is able to overrule the subordinate if necessary. Clause 66 is often better than Clause 2 in this case, unless further time is required before having to consider arbitration. Clause 2 may be difficult to employ when subordinates' decisions come out bearing a rubber stamp signature of the engineer or his name written at the bottom of the letter as if signed by him. It is also essential to ensure that the Clause 66 decision is that of the engineer and not that of a subordinate similarly signed.

Some engineers prefer to give a Clause 66 decision straight away as they consider they may then be more impartial.

A Clause 66 decision may be requested by the contractor to loosen the deadlock which may occur when the auditors, particularly of another authority, dispute with the engineer one or more of his decisions; or when a purchasing authority not named in the contract is in dispute with an agent authority, named in the contract as the 'Employer', as to the responsibility for the payment to the contractor for part of the work certified. Such a case might be when the responsibility for problem design work is being debated.

Sometimes, one has occasion to wonder what sub-conscious effect auditors have on engineers' 'non-clause 66 decisions' through their disputing some of the engineers' past decisions. Engineers' subordinates have been known to ask for auditors' comment before making decisions.

It is probable that the 5th Edition of the ICE Contract in its various altered clauses, has restored to engineers some of the power which had been whittled away in the past. This power is essential for the smooth working of the contract.

If it were shown that the employer or his auditors had influenced an engineer's Clause 66 decision, it could be contended that they were in breach of contract. If a third party authority, or their auditors, had influenced an engineer's decision, the engineer might be guilty of misconduct. Discussion of this and action to be taken is outside the terms of this paper but is dealt with more fully elsewhere.[5]

Engineers must be constantly on watch to ensure that they do not abrogate their powers to auditors or the employer's other principal officers. Their decisions must be their decision alone. They must appear to the parties to be impartial as well as being impartial.

It is a matter of concern that contractors often doubt the efficiency of the present system of contracting and consider there should be a full (or part) time arbitrator on each contract to settle disputes and differences that arise. The engineer's powers would, however, be reduced to the extent that the whole system of contracting would have to be changed.

On the present basis of contracting the contractor and the employer have, as their safeguard Clause 66 decisions, with arbitration in the background.

Where the named engineer is known to be unreasonable before tenders are submitted, tender prices should allow for this since all cannot be retrieved by Clause 66 protection. It is unfortunate, however, that sometimes engineers are changed during a contract. This may be classified as a contractual risk.

An engineer's failure to administer the contract is considered in detail by Abrahamson[5].

An arbitrator put forward by one party in accordance with Clause 66, is frequently turned down on principle by the other and vice versa, unless the name has been produced by an independent third party or the work is of a specialist nature. A Presidential nomination is frequently used.

Clause 66(5)(c) in the January 1986 reprint states that any reference to arbitration may, unless the parties otherwise agree in writing, proceed notwithstanding that the works are not then complete, or alleged to be complete. However, due to the time delays, most contracts will be finished by the time the arbitration takes place. The arbitration can, however, take place earlier under this revised clause.

Some authorities continue to delete clause 66 forcing the contractor to go to the Courts for his protection. Such an alteration is normally against a contractor's interest, whose case is generally better dealt with by an experienced and capable arbitrator at a pre decided date for a hearing in private without the formality and inconvenience of a Court, but with the back up of the High Court, and without the problem of the Court's powers described earlier.

Arbitration and litigation are complementary. They are not competitors. Where there is an arbitration clause, Court proceedings may be stayed for arbitration as previously described. The 1983 ICE Arbitration Procedure makes some provision for concurrent hearings. Where several parties are involved the Courts may have to be used.

Both with arbitration and litigation reluctant respondents may cause considerable delays, but a good arbitrator may restrict this.

Clause 67 — Application to Scotland
If the works are situated in Scotland, useful information on Scots law was given in the old ICE Arbitration Procedure (1973) and by Inglis[10].

Clause 72 ON — Special conditions
These should be particularly noted and studied.

5. 4th Edition of the ICE Contract

The General Conditions of Contract and Forms of Tender, Agreement and Bond for use in connection with Works of Civil Engineering Construction, Fourth Edition, approved and recommended for general use by the Institution of Civil Engineers, the Association of Consulting Engineers and the Federation of Civil Engineering Contractors, were first issued in January, 1955. Subsequent additions were made as follows:

Metrication	January, 1969
Tax fluctuations	January, 1970
Value Added Tax	1 September 1971 revised 23 March, 1973

Available as a loose-leaf clause for use in appropriate cases is the Contract Price Fluctuations Clause (first issued on 29 March, 1973 and revised in June, 1973) issued to replace the Variation of Price (Labour and Materials) Clause.

This edition in regard to claims is increasingly rare, but some employers still use it, so the 4th Edition cannot yet be filed away. When using the 4th Edition, one must think 4th Edition. Memory the clauses will fade with time, so it is essential to re-read and re-study them.

The general numbering of clauses between the 4th and 5th Editions varies little, but the detail is considerably less in the 4th Edition. This is illustrated by the number of pages for the first 68 clauses, 29 in the 5th Edition against 18 in the 4th Edition. A certain 'custom' had grown up for dealing with items which are now covered in the 5th Edition.

Claims for entitlement were dealt with in fewer places and more briefly in the 4th Edition, yet general agreement was normally reached, possibly because auditors queried engineer's decisions less often and compromise was easier.

A number of changes were forced by time. Interest, for example, was rarely reclaimable under the 4th Edition and was low — Clause 60(3). Interest did not even feature in the Index to General Conditions. High interest rates did not normally exist then.

Most people now prefer the 5th Edition and have become so used to it that they tend to forget the clauses of the 4th Edition. Nevertheless, on 4th Edition contracts the 4th Edition must be used and one cannot turn a blind eye to it or forget it and use a different edition or contract.

One must, however, bear in mind the 'custom' referred to above which had developed in its use. This, to some extent, became an inferred term of 4th Edition work, though such 'custom' will be gradually forgotten and, no doubt, soon be disputed.

6. Sub-contractors

INTRODUCTION

Nominated sub-contractors are provided for in the main contract and if their form of sub-contract with the main contractor is based on the Form of Sub-Contract designed for use in conjunction with the ICE General Conditions of Contract issued by the Federation of Civil Engineering Contractors, they have a reasonable amount of protection, particularly over disputes. There is always the risk of the contractor going out of business and 5th Edition Clause 59C after the sub-paragraph (b) only states the employer 'shall be entitled to pay ...' and not 'shall pay'.

Nominated sub-contractors supplying and fixing plant are often employed on the IMechE/IEE/Association of Consulting Engineers Model Form of General Conditions of Contract A for use in connection with home contracts with erection. There may be differences between this contract and the main contract, particularly regarding damages. The main contractor must be alert to this and object to the nomination where necessary, until he is covered — see the section under Clauses 58/59A/59B/59C.

With domestic (non-nominated) sub-contractors on the form of sub-contract referred to above, there is less protection, but the contractor and sub-contractor are often used to working together and the arrangement seems to work reasonably well.

Other forms of sub-contract are often used, particularly with domestic sub-contractors working for national firms; these firms often use their own form of sub-contract. Some of these sub-contract forms are very one-sided and could be dangerous to the sub-contractor if implemented. However, when trouble occurs common sense and horse trading, without regard to the contract, often settle the dispute as both parties are generally anxious for a settlement.

The sub-contract order is normally the acceptance of the tender. However, some sub-contract orders require acknowledgement. Examining the small print shows the sub-contract order to be only an offer for acceptance by the sub-contractor by acknowledgement. Such an order may not be the same as the sub-contractor's tender and by his acknowledgement he has frequently committed himself to conditions he did not appreciate. These may work very harshly against him. He is ill-advised to acknowledge without studying the small print and the order (offer).

The sub-contract documents and covering letters require careful study when a claim is being made. When the claim for entitlement is basically against the main contractor, rather than the employer, a counterclaim is normally to be expected.

THE CONTRACTOR CAUGHT IN THE MIDDLE

Too often the main contractor finds himself caught between the employer's requirements and those of the sub-contractor which do not tie up contractually. This is often the main contractor's fault. For example, the main contract may require the main contractor to include for running sand whereas the sub-contractor has excluded running sand from his quotation.

In the case of nominated sub-contractors, the main contractor could have failed to object to the nomination under Clause 59A of the 5th Edition because he either failed to notice the difference, did not appreciate the difference, or was in a hurry to get the job moving. This latter is particularly dangerous in the situation where the 'paperwork' follows.

With domestic sub-contractors, the main contractor often bases his quotation on some sub-contractor's tender rates. When the time comes to employ them they may claim to be too busy, or that the main contractor is out of time, or the engineer will not accept them or their goods or their offer generally. The contractor is left to finance the difference with another sub-contractor out of his risk and profit item. There is no easy solution for the contractor over this, other than to be careful of the sub-contract rates he feeds into the bill, particularly where they are excessively low.

The main contractor's immediate defence to a claim from a sub-contractor is to counterclaim against the sub-contractor to at least the same amount and to withhold money. Where responsibility for the sub-contractor's claim rests with the employer, the main contractor will normally be helpful by offering to let the sub-contractor join him in arbitration, often claiming more money for himself as well.

The result is that claims where the employer is responsible tend to be settled contractually, whereas claims where the main contractor is responsible are 'horse traded'. Where the claim is not dealt with satisfactorily by the main contractor, the sub-contractor tends to avoid future work with him and this has a powerful influence in the horse trading process.

The default of a sub-contractor may cause considerable cost and delay to a main contractor whatever the reason. When a nominated sub-contract fails, for whatever reason, a new nomination must be requested. It is unwise normally to let a nominated sub-contractor on the site without a binding contract. With domestic sub-contractors this will depend on the relationship and size of sub-contract and normal methods of ordering but to do so often leads to trouble and dispute.

Letters of intent can be dangerous unless very carefully prepared. The main contractor may be caught with a useless letter which binds no one, or a partial contract which leaves him unprotected on further negotiation. There must be no loopholes and the letter must be specific on the liability of both parties.

DELAYS
The sub-contractor may be delayed in starting and carrying out the work by the main contractor, or his other sub-contractors, or because of the employer.

When the responsibility is that of the employer, the problem is not great for the reasons given above. Where the responsibility rests with the main contractor or one of his other sub-contractors, the sub-contractor may be able to horse trade as described, or even not start the work without extra money.

Equally, the sub-contractor may delay the main contractor and the other sub-contractors by starting late or proceeding slowly. What happens then depends very much on the circumstances.

A sub-contractor's problem is often that he is unable to find out the reasons for the delays until discovery, and the 1983 ICE Arbitration Procedure may help in this respect.

ARBITRATION/COURTS
The amount of money at stake is normally less in the case of sub-contractors, but the firms are smaller and feel the blow just the same.

There is a tendency for these cases, particularly with domestic sub-contractors, to be settled by writs in the courts than by arbitration, or threatened arbitration. The claims are often more straightforward and less involved and concern finance more than anything else. Sometimes, particularly with nominated sub-contractors, the employer becomes involved and may be asked to adjudicate. A contractor may run into trouble where a main contract is sealed and a sub-contract is 'under hand' due to different time bars.

Arbitration, where it does take place, tends to be more on Clause 66 of the main contract than on the 'Disputes Clause' of the sub-contract. This is to be expected from what has been said before.

MANAGEMENT SUB-CONTRACTS
Management contractors sometimes use the sub-contract form for their sub-contractors. The form was not designed for this purpose and gives inadequate protection to these sub-contractors. A main contract form is safer for them.

Those asked to tender on a management sub-contract basis should ensure that the management contractor is guaranteed by a company or organisation of repute. Otherwise, it is possible for a firm to be created for a specific project as a management contractor and be liquidated at a suitable time.

PRICING A BILL ITEM FOR DELAY
Some bills of quantities contain an item to price for delay per week as the employer's responsibility. A rough guide to the reasonableness of an amount inserted by the contractor for a week's delay is obtained by taking the tender sum, deducting the materials and nominated content, dividing by the number of working weeks programmed and adding a factor for inflation to allow for the work being done after the end of the programmed period.

Such an item is unsatisfactory. The assessment produced above may be wide of the mark and so many other factors are involved, both in the buildup of the bill items, and during the phases of the contract.

7. References

1. BURKE, H.T. (1976) Claims and the standard form of building contract. Chartered Institute of Building. pp43

2. HARLOW, P.A. (1986) Contractual claims: an annotated bibliography. Chartered Institute of Building. 5th edition.

3. WOOD, R.D. (1986) Builders' claims under the JCT63 form of contract. Chartered Institute of Building. pp61

4. DUNCAN WALLACE, I.N. (1970) Hudson's building and engineering contracts. Sweet and Maxwell. pp921 plus first supplement 1979.

5. ABRAHAMSON, M.W. (1979) Engineering law and ICE contracts. Applied Science Publishers. 4th edition. pp485.

6. INSTITUTION OF CIVIL ENGINEERS. (1983) Arbitration procedure England and Wales. pp4. With a separate version for Scotland.

7. WALTON, A. AND VITORIA, M. (1982) Russell on arbitration. Stevens and Sons. pp602.

8. PARRIS, J. (1974) The law and practice of arbitration. George Godwin. pp145

9. Arbitration Act 1950. (1979) HMSO.

10. INGLIS, I.G. (1977) Some points of difference between Scots and English laws of arbitration. *Arbitration 43* (2), pp72-78.

11. MUSTILL, SIR MICHAEL J. AND BOYD S.C. (1982) The law and practice of commercial arbitration. Butterworth.

12. HANKER, G., UFF, J., AND TIMMS, C. (1986) The Institution of Civil Engineers. Arbitration Practice. Thomas Telford. pp244.

13. BARNES, M. (1977) Measurement in contract control. Institution of Civil Engineers. pp296

14. BARNES, M. (1986) The CESMM2 Handbook. Thomas Telford. pp319

15. INSTITUTION OF CIVIL ENGINEERS (1982) Conditions of Contract Standing Joint Committee (CCSJC) Guidance Note (GN) 4 — Interest on overdue payments (Ref. CCSJC/GN4/April 1982). pp1

8. Appendices

APPENDIX I. EXAMPLE OF DETERMINING THE AMOUNT BUILT INTO A TENDER FOR PRELIMINARIES, OVERHEADS AND PROFIT

CONTRACTORS METHODS OF PRICING TENDERS

With large and with specialist firms, the bills of quantities may be priced first on the net cost of labour, material and plant directly attributable to the items concerned. An addition is then made to the basic net rates, as may be required, to cover such of the following as are not built into the priced parts of the bill:

site overheads and preliminaries;
firm's overheads;
risk, profit and financing;
economies and better buying.

With such firms, monitoring on overheads frequently takes place as the job progresses.

With other contractors, bills of quantities tend to be priced with the main measured items first built up from the basic net labour, plant and material elements, or from domestic sub-contract quotes, to which about 1/5th or 1/6th is added as the contribution for preliminaries, overheads and profit. The majority of the rest of the preliminaries, overheads and profit amount is then distributed between the bill preliminary items, PC sum additions and the balancing item after allowing for economies and better buying.

Some contractors add a higher percentage to labour only, distributing the rest of the preliminaries, overheads and profit as above. Some use a lower percentage on domestic sub-contract work.

BREAKING DOWN A TENDER TO FIND THE AMOUNT ADDED FOR PRELIMINARIES, OVERHEADS AND PROFIT

It is normally possible for an experienced eye to break down a priced bill to determine the amount contained therein for preliminaries, overheads and profit. The amount is ascertained as follows:

(a) ***Measured work element***

If inspection indicates that the addition on the measured work for preliminaries, overheads and profit is 1/5th, and if the measured work in the bill is £240,000, then the addition to be extracted here is 1/6th of that amount, ie, £40,000. In addition, when work is short, there may be a further amount added from intended better buying of the materials and the placing of domestic sub-contracts, and from intended greater efficiency with labour and plant, particularly intending a reduced length of job. This may amount to £20,000.

(b) ***Provisional and PC sums***

The preliminaries, overhead and profit element included in the provisional and PC sum parts of the bill of quantities will normally be apparent on inspection of the bill items. Assume that in the tender under examination these amount to £2,000.

(c) ***Contingencies and daywork items***

A contribution from contingency and daywork items towards preliminaries, overheads and profit may or may not have been allowed depending on the wording of the bills and the contractor's method of pricing. Assume that in the tender under examination £1,000 was allowed.

(d) ***Preliminaries and balancing item***

Assume that the preliminaries amounts entered in the bill amount to £25,000 and the balancing item is minus £2,248.

(e) ***Summation of amounts entered***

Summarising the amounts assessed above for the preliminaries, overheads and profit in the bill gives the sum of £85,752 as included in a tender amounting to, say, £360,000.

BREAKING DOWN THE PRELIMINARIES, OVERHEADS AND PROFIT AMOUNT

The amount of £85,752 is the amount the contractor added to cover:

site overheads and preliminaries;
firm's overheads;
risk, profit and financing.

Site overheads and preliminaries

Let the site overheads and preliminaries, when calculated out per week in a way similar to the example in Appendix II but for the programmed period, amount to the sum of £35,752.

Firm's overheads
Firm's overheads vary with the size of firm, the type of work and the allocation of supervisory personnel to firm's and site overheads.

As a rule, the addition for medium sized firms is 8% to 13% of turnover, with bigger firms going lower and specialist sub-contractors higher to about 25% of turnover. The reason that larger firms have a lower firm's overhead is largely due to the type of work they do and to their normally higher preliminaries and site overhead amount.

Let the firm's overhead addition in this case amount to £32,000 this being for the programmed period, not the contract period.

Risk, profit and financing
By deduction, the amount added for risk, profit and financing is £18,000. There is, however, a risk item built into the labour element of each rate to cover the contractor against abnormal weather and other problem areas, including particularly that of labour. The cost of the problems brought about by events for which the employer carries responsibility are not covered in a normal tender.

SUMMARY OF THE PRELIMINARIES, OVERHEADS AND PROFIT ITEMS
In this example, the additions for preliminaries, overheads and profit added at tender stage is assessed as £85,752 broken down as follows:

	£
Firm's overheads	32,000
Site overheads and preliminaries	35,752
Risk, profit and financing	18,000
	£85,752

APPENDIX II. EXAMPLE OF A TYPICAL QUANTIFICATION

Every quantification has to be varied to suit the claim, but the following summary and build-ups will give some guidance. The delay in this example is taken as 54 working weeks from the programmed completion date, with 50 of these weeks being the responsibility of the employer.

SUMMARY OF CLAIM FOR ENTITLEMENT TO REIMBURSEMENT OF LOSS AND EXPENSE DUE TO THE DELAY, DISRUPTION AND OTHER FACTORS
Quantification of the cost of the disruption and of the delay of 50 working weeks on the critical path, which is the responsibility of the employer, ie, 54 working weeks, less four weeks for the weather in the period ending 31 May 1984 is as follows:

	£
Additional site overheads	
(a) *after the programmed period*	
38 weeks @ £4,905.60 per week	186,412.80
12 weeks @ £3,386.60 per week	40,639.20
(b) *during the programmed period*	
15 weeks @ £600 per week (see below)	9,000.00
Additional preliminaries (see below)	2,280.00
Additional firm's overheads 50 weeks @ £755.20 per week (see below)	37,760.00
Adjustment for varied work content (see below)	XXX.XX
Additional plant cost (say) (see below)	15,000.00
Additional labour cost (see below)	15,000.00
Add for loss of contribution due to disruption (see below)	20,000.00
Additional ...... cost (see below)	1,000.00

Items not agreed in the final account (see below)	10,000.00
Increased costs or additional increased costs (see below)	5,000.00
Add for extra cost of financing and gross profit (see below)	XX,XXX.XX
Additional financing cost of the retention (see below)	1,000.00
TOTAL BASIC CLAIM FOR ENTITLEMENT	£XX XXX.XX

PLUS INTEREST CALCULATED IN ACCORDANCE WITH THE CONTRACT
(see below)

ADDITIONAL SITE OVERHEADS

(a) *After the programmed period*

Item	Basic cost per week (£)	
	date to date	date to date
Agent and car (Note 1 Page -)	500.00	500.00
Engineer(s) and car (Note 2 Page -)	400.00	
Visiting CM and car 15% of £800	120.00	120.00
Visiting QS and car 75% of £400	300.00	300.00
Visiting storeman 50% of £180	90.00	
Visiting clerks 50% of £180	90.00	
General foremen	250.00	250.00
Foreman (part non-working) 2 No. x 50% of £220	220.00	
Site labourers/chainman (2)	350.00	175.00
Visiting maintenance fitters 1/5	90.00	45.00
Site materials Van 1/5	70.00	20.00
Watchmen	200.00	200.00
Welfare	50.00	30.00
Safety	50.00	30.00
Huts	100.00	70.00
Services	120.00	100.00
Cleaning site	80.00	80.00
2 No. site vans (transport)	400.00	400.00
Site plant	400.00	300.00
Small tools	100.00	80.00
Temporary works and maintenance of roads	100.00	50.00
Small materials	50.00	20.00
Tests	50.00	
Sundries, surveying equipment and materials for setting out	200.00	100.00
TOTAL PER WEEK	£4,380.00	£2,870.00

Allowing for inflation increased amounts (when no provision is made in the contract for adjustment) the costs per week will be:

38 weeks @ £4,380.00 x 1.12 = £4,905.60 per week
12 weeks @ £2,870.00 x 1.18 = £3,386.60 per week

It may be preferred to use actual rates or to apply the Baxter Factor. Each rate requires to be built up in notes and so indicated against the item as for the first two items above.

(b) *During the contract period*

XX weeks of:		£ per week
GF (intended to be working half time)		125.00
Visiting QS (75% instead of the 25% intended)		200.00
etc.		XXX.00
	say	600.00

for XX weeks = £9,000.00 extra

The amounts above are carried to the 'Summary of claim for entitlement to reimbursement'.

NB. In order to demonstrate a large number of items, these site costs do not tie in with the example in Appendix I. The job shown in Appendix I could not economically carry weekly site costs of this magnitude.

ADDITIONAL PRELIMINARIES

Preliminaries Item			£
1/1	£2,000 / No. of weeks	x delay in weeks	XXX.00
1/2	£1,000 / No. of weeks	x delay in weeks	XXX.00
1/3	Lump sum of ? % of £500		XXX.00
1/4	Cost of ? after?		XXX.00
1/20	etc. (Insurance)		XXX.00
1/25	etc. (Bond)		XXX.00
			£2,000.00

Allowing for inflation increased amounts (when no provision is made in the contract for adjustment) the total will be £2,280.00.

This figure is carried to the 'Summary of claim for entitlement to reimbursement'.

NB. Where the preliminary items are not realistic figures, they should be costed above. This head of claim is particularly suitable where the 'Method of Measurement for Road and Bridge Works' is in use.

ADDITIONAL FIRM'S OVERHEADS

The amount required at the time of tender for firm's overheads on a contract which ran according to programme was 8.89% on gross turnover (see auditor's certificate/calculations from the balance sheet).

The amount of firm's overheads to be recouped per week in the 50 working weeks originally programmed was thus:

$$\frac{8.89\% \text{ of } £360{,}000}{50 \text{ working weeks}} = £640.00 \text{ per week}$$

For the 50 working weeks delay for which the employer is responsible, this figure has to be increased by 18% inflation making £755.20 per week for 50 weeks = £37,760.00.

This figure is carried to the 'Summary of claim for entitlement to reimbursement'.

NB. The method of arriving at the inflation percentage should be given. The inflation amount may be from the mid point of the original programmed period to the mid point of the extra period resulting from the delay and other factors for which the employer carries responsibility. Overheads go up with inflation, and it is a fallacy to represent, as some do, that inflation is covered by the method without the inflation factor addition.

ADJUSTMENT FOR VARIED WORK CONTENT
Credit or debit on the items above resulting from the excess or reduced amount of work from that calculated in the original bill should be computed. This sum is based on the percentage addition on the basic measured items or additional work as appropriate and is carried to the 'Summary of claim for entitlement to reimbursement'.

ADDITIONAL PLANT COSTS
The extra plant costs due to disorganisation, disruption and out of sequence working, for which the employer is responsible, amounts to the sum of £XX arrived as explained below.

Annexure 1 gives the programmed plant (not reproduced here)

Annexure 2 gives the plant actually used (not reproduced here)

The difference adjusted for the fuel used with an adjustment for the extra work ordered, multiplied by the factor;

$$\frac{\text{(weeks delay for which the employer is responsible)}}{\text{(Total weeks delay)}}$$

is carried to the 'Summary of claim for entitlement to reimbursement'.

NB1 It may be more convenient to include drivers hired with the plant here and possibly all drivers and an adjustment in the next paragraph might be necessary.

NB2 When unfair methods are used by engineer's subordinates to calculate plant costs to a contractor, the matter is best referred to the engineer for his comment.

NB3 It may be possible to assess the additional plant cost from the plant plus drivers and fuel on the site for the respective extra weeks of the delay.

EXTRA LABOUR COSTS
The extra labour costs due to the disorganisation, disruption and out of sequence working, for which the employer is responsible, including subsistence and travelling but excluding the cost of plant drivers included under 'Additional Plant Costs' above (see note NB1 above) amounts to the sum of £XXX arrived at as follows:

	£
Total labour cost including subsistence and travelling but excluding plant drivers included above	XXXX
Less labour used on extra measure (including/excluding (as appropriate) plant drivers)	XXXX
	XXXX
Less labour included in bills (including/ excluding (as appropriate) plant drivers)	XXXX
Total extra labour cost	£XXXX

This figure then requires to be multipled by a factor to allow for any delay for which the employer does not carry responsibility and the result carried to the 'Summary of claim for entitlement to reimbursement'.

Alternatively, it may be possible to assess the extra labour for the respective extra weeks of the delay for which the employer carries responsibility from the extra gang weeks, site labour and tradesmen weeks involved.

LOSS OF CONTRIBUTION DUE TO DISRUPTION
The contribution to overheads and profit from the intended greater efficiency and better buying on the contract, when placed, was never achieved. That failure to achieve was due to disruption, disorganisation and out of sequence working caused. The loss of contribution intended from this greater efficiency and better buying was £XX,XXX.XX arrived at as follows:

	£	
Tender sum	360,000.000	
Deduct	XXX,XXX.XX	(see below)
Leaving	XXX,XXX.XX	
10% of XXX,XXX.XX	£XX,XXX.XX	(to page ...)

The £XXX,XXX.XX was in the tender as follows:

Prelim's Section
Daywork
Prov Sums
PC Sums
P&A on PC Sums
Traffic
Dewatering
Contingencies

£XXX,XXX.XX

The 10% was the intended amount at the time of the tender.

Subsequent Paragraphs for **ADDITIONAL .... COSTS**

Paragraphs such as increased costs of moving material or other reasons may be appropriate and the figures carried to the 'Summary of claim for entitlement to reimbursement'.

ITEMS NOT AGREED IN THE FINAL ACCOUNT
These items should be listed.

INCREASED COSTS OR ADDITIONAL INCREASED COSTS
This item will depend on the contract and the items in the previous paragraphs and may apply only to the additional increased costs on materials. The amount claimed is carried to the 'Summary of claim for entitlement to reimbursement'.

ADDITIONAL COST OF FINANCING AND GROSS PROFIT
(The contractor is entitled to a reasonable percentage addition in respect of profit on certain clauses (including Clause 12) and additional financing is a cost for reimbursement separate from interest (see below).

Addition due on:

Item	£
XXXX	XXX
XXXX	XXX
	£XXX

at the rate of 10% (or other percentage) this is £XXXX.

This figure is carried to the 'Summary of claim for entitlement to reimbursement'.

ADDITIONAL FINANCING COST OF THE RETENTION
The retention of £XXX has had to be financed for an additional 53 weeks (allowing for 3 weeks holiday) at a cost of:

$$\text{£XXX} \times \frac{53}{52} \times \frac{14}{100} \text{ (or other percentage)}$$

$$= \text{£XXXX}$$

The figure is carried to the 'Summary of claim for entitlement to reimbursement'

It may be necessary to do this calculation in greater detail using the interest rates paid by the contractor to the bank for the number of days concerned, multiplied by the financial amount actually out over anticipated, divided by the average number of days per year in the period.

(Other amounts may also be involved)

PLUS INTEREST
See note under Clause 60 in the text.

Interest cannot normally be calculated in advance of settlement due to interest rates changing and to the delays in negotiation. A schedule of interest rate changes should be given together with a chart of dates.

CLAIM
Summarise the claim.

The claim is for an extension of time of 54 weeks as detailed on page ... (which will normally summarise contract week numbers, weeks delay, reason and clauses) and for reimbursement of the cost of £XXX plus interest as given above. The engineer's certification is awaited.

(Clauses under which the claim is made should be quoted above or in the text, as appropriate, and should not tie the engineer down unnecessarily by being too specific).